Insects

photographs
and text by
MAVIS LESSITER

HERITAGE OF
NEW ZEALAND

THE BUSH PRESS
OF AUCKLAND

CONTENTS

Introduction	3
Classifying Insects	4
Praying Mantis	6, 7
Stick Insects	8, 9
Crickets, Katydids and Grasshoppers	10, 11
Weta	12, 13
Cicadas	14
Beetles	15, 16, 17
Weevils	18
Cockroaches	19
Bugs	20, 21, 22
Earwigs, Ants	23
Bees	24, 25
Wasps	26, 27
Insect Study	28, 29
Flies	30, 31
Index	32

MATING MANTIDS
The larger female often devours the smaller male after mating. The large abdomen is swollen with eggs. These are South African mantids.

INTERESTING INSECTS

There are more than 20,000 different kinds of insects to be found in New Zealand. Many are rare or found only in small areas. This collection features the more common groups with examples photographed about the house and garden.

Insects are not all pests. In fact some have been specially introduced to New Zealand because of the good they do. Both "good" and "bad" can be found in the garden.

The pests are those which feed on food, plants and materials used by people. The useful insects, like bees, may spread pollen from flower to flower and so help fruit to develop. Others feed on garden pests and help reduce them, like the ladybird on aphids.

All the insects in this book are "interesting" in some special way. The close-up pictures will help to identify the common insects of the garden and reveal the beauty of these creatures.

THE NATURE OF INSECTS

The body of an adult insect consists of three parts: head, thorax and abdomen. On the head are the eyes, antennae (or feelers) and mouthparts. The legs (and wings in the winged species) are attached to the thorax. The abdomen contains the digestive and reproductive organs and the excretory system.

Insects breathe through small port-holes or spiracles. These are located along each side of the body and draw air, through tubes called tracheae, to all parts of the body.

Insect species differ in their feeding habits, and so are equipped with different types of mouthparts. Some insects have strong jaws for chewing; others have a sharp tube for piercing plants, or other insects, and sucking out the juices.

The skin (cuticle) is made of chitin, a material that is both flexible and very strong. It can be hard or soft depending on the species.

The life cycle of an insect varies from species to species but the basic pattern is the same. After mating, the female generally searches for a suitable food plant or the correct habitat and lays her eggs on or near the future food source. After a period for incubation the egg hatches, and the larva or nymph emerges. The larva is a form of grub or caterpillar. A nymph is an immature insect shape. As the young insect feeds and grows, it will shed its outer skin several times. When it is fully grown it will either pupate and develop inside the pupal case until it emerges as an adult insect, or as in some other groups, the nymph becomes an adult insect at the final moult, without going through the pupal stage. This process may take as little time as a few days or be spread over several years.

CLASSIFYING INSECTS

World-wide there are hundreds of thousands of different kinds of insects. To keep them in some sort of order a Swedish scientist named Linnaeus devised a means of classification that uses two Latin names. The first is the Genus or generic name which tells us which insects are alike and signifies their close relationship. The second name, which is the species or specific name, tells us which of the insects it is. This is the basis of classification.

All the closely related Genera (singular genus), that is those with the same general pattern, are grouped into a Family. Related Families are then grouped together into an Order. All the related Orders, in turn, form a Class. The class for insects is Insecta.

The Latin language is used as an international standard for naming insects, just as it is also used for plants and other animals. This system of classification is just like a family tree, or a chart showing relationships. The insects often have names by which they are commonly known by the local people, but the Latin name is constant around the world.

To decide the Genus an insect belongs to can be quite complicated. Scientists have to look very carefully at the insect. Sometimes the difference is very obvious but it may be as slight as how a tiny hair is placed, or even the structure of one of the internal organs. There have been occasions when an insect was named and then, years later, new technology has shown minute differences between those previously thought to be identical. When this happens the insects are renamed. Classifications are revised and perfected, as scientists reach a better understanding of how different groups relate to one another.

Figure 1 is an example of how this classification or "tree" works. When a new species is found, and described for the records, it is first grouped with its relatives and given a generic name. A specific name is then given and may indicate some aspect of anatomy or colour; or it may be named after whoever discovered it, or where it was found.

Fig. 1 INVERTEBRATA (invertebrates) Animals without backbones

PHYLUM	—	Arthropoda	Jointed-legged invertebrates
CLASS	—	Insecta	Adult (insects) have three pairs of legs
ORDER	—	Diptera	2-winged insects (flies)
FAMILY	—	Muscidae	House flies and their close relatives
GENUS	—	Musca	The house fly group
SPECIES	—	domestica	THE HOUSE FLY

This is how the House Fly is classified and given the scientific name Musca domestica.

COMMON PRAYING MANTIS

The Common Mantis occurs in both North and South Islands.

PRAYING MANTIS

Beautiful and elegant, the praying mantis is one of the most aggressive killers in the garden. It will capture, kill and eat wasps, flies, blow flies, cicadas, crickets, moths, in fact, just about anything in the insect world, even other mantids.

There are two species in New Zealand. The Common Mantis, *Orthodera ministralis* is found in both the North and South Islands. The South African Mantis *Miomantis caffra* was first recorded in the Auckland region in 1978. It has now gradually spread and can be found from Paeroa to Kaitaia.

The Common Mantis is green with a bright spot of coloured purple pigment on the inside of the front leg.

The South African female mantis is much larger than the Common Mantis, has a more slender neck and, at the base of the "throat" a necklet of gold spots. The mantids have been found in different colours, ranging from dark orange to pale straw, as well as the more usual green. Both males and females have been found with this colour range, but the males are more slender, and have clear iridescent wings.

SOUTH AFRICAN PRAYING MANTIS
Miomantis caffra 60mm

These mantids are normally green, though there are colour variants.

Mantids lay their eggs in an egg case or "ootheca". The Common Mantis builds an egg case like a tiny Ayers Rock, while the South African Mantis makes her egg case flatter, oval and with a stripe along the top edges that ends in a little tail. In both species the egg cases contain a large number of chambers and in each is an egg which develops into a tiny mantis nymph.

When they hatch the nymphs can be very active. They quickly disperse and start to feed on aphids and other small insects, including other mantis nymphs. They moult several times but their wings are not fully developed until the last moult.

Both species have long forelegs with large spines. These are used to catch and hold prey. The mantids clean their forelegs carefully after each meal. The forelegs are very long, and when the mantis is hunting the legs are folded, which is why the mantis is called "Praying Mantis".

Maori class praying mantids with stick insects, because both groups have wide-set legs which suspend the body in the air like a twig (this is a perfect camouflage). The names vary according to district and include Ro, Whe and Wairaka.

MANTIS EGG CASE

The little tail is the way to know this egg case was made by a South African Mantis.

STICK INSECTS

Stick insects feed on a variety of trees and plants and are frequently found on manuka. They have perfect camouflage, lying along the branch or leaf where their colour and twiggy appearance make them almost invisible, unless they move. New Zealand stick insects have no wings, unlike their overseas relatives.

Two common types of stick insect are the "smooth" and "knobbly" stick insects. Their colours range from bright leaf green to dark greyish-brown.

An unusual ability of these insects is for the unfertilised females to lay eggs that will hatch, but, all the progeny will be female. When the eggs of mated females hatch, the nymphs can be either male or female.

These insects are very easy to keep as pets. They will live quite happily in a light, airy container. Give them plenty of fresh food, water in a shallow container and an occasional mist spray. For food try the leaves of rimu, manuka, pohutukawa and roses. The eggs are dropped to the ground so a careful search through the droppings could reveal barrel-shaped eggs which have intricate patterns. These will hatch if they are stored on absorbent paper in a small container. The hatched nymphs are miniatures of the adults. It is important to handle stick insects with care; their long legs are very fragile and may drop off if they are damaged.

Maori class stick insects with praying mantids as Whe or Ro or Wairaka.

STICK INSECTS 75-150mm

Stick insects hold onto a food plant with small hooks on their feet. The stick insect sometimes rests with the two front legs extended in front of the head. In this position, as long as it remains motionless, it blends in with the twigs and branches; a perfect camouflage. If a leg is lost, the limb can grow again.

CRICKETS, KATYDIDS and GRASSHOPPERS

All these insects are frequently found in the garden, where they feed on grasses and other plants. On warm evenings crickets and katydids create the chorus that we associate with summer. The sound the male insects make is their mating call.

These insects lay their eggs on a food plant or in the soil. The young that hatch are replicas of the adult, but without wings. They moult several times as they grow, and the winged species develop wing buds. After the last moult they are fully developed and the "singing" or mating call of the males begins.

The Katydid and Green Grasshopper are winged and are known as Longhorned Grasshoppers. The true New Zealand grasshoppers have short antennae and (with the exception of one species) have only partly-developed wings and cannot fly. These are mostly found in the mountains.

Maori names for crickets include Rirerire, Kakarapiha, Pihapihareina and Pihareinga. In historical records the katydid is called Kikipounamu. Names for grasshoppers vary according to location. They may be called Mawhitiwhiti, Mowhitiwhiti, Rawhitiwhiti, Kauwhitiwhiti and Koeke.

LOCUST

Locusta migratoria 45mm

The locust is a large grasshopper. It flies in a swift erratic manner, making a clapping sound as it does so. In some countries the locusts congregate in large swarms, moving through the countryside causing complete devastation. Fortunately in New Zealand this does not happen.

GRASSHOPPER
Conocephalus spp.

This small Longhorn Grasshopper feeds on grasses.

KATYDID
Caedicia simplex 75mm

During the day the Katydid, a Longhorn Grasshopper, is found on shrubs and flower plants. The female has a short, broad ovipositor and a very quiet song. The male Katydid sings more loudly, adding to the summer-evening chorus. Though the Katydid has large wings, it is a rather sluggish flyer. It is a nocturnal insect and often attracted to lights at night.

BLACK FIELD CRICKET
Teleogryllus commodus 25mm

Often an unwelcome visitor to the unguarded picnic basket, crickets also enter buildings. While they may be a bit of a nuisance and cause considerable damage to grassland, they do not bite or sting. The mating call of the male is a cheery sound on summer evenings.

AUCKLAND WETA
Hemideina thoracica
40-50mm

The male has an elongated black head and the female has a small round head and a long ovipositor. The small hole, on both sides of the shank of the front legs is the ear, and the portholes or spiracles along the body are for taking in air. The long heavy-spined back legs are raised over the body in an "attack" position if the weta is threatened.

WETA

Wetas are found only in New Zealand and include some of our largest insects. There are a number of species. They range in size from 20mm to 100mm. The largest species is *Deinacrida heteracantha*. This weta is only found on Little Barrier Island. Recently a small colony of another giant weta has been found in the North Island where they are being carefully protected.

Each species has its own preferred habitat, including rotting logs, banks, caves, under rocks, in burrows and in the roofs and walls of old buildings. Wetas feed on vegetables and plants, fruit and foliage both fresh and decaying, and on other insects.

Some of the wetas are stong heavy-bodied insects while others, like the cave weta, are delicate dainty creatures. Common features are very long antennae and long legs for jumping. In spite of their fierce appearance, they will usually only bite if handled roughly. Strong spines on the hind legs can inflict a sharp scratch if not treated with respect.

WETA
BUSH WETA (Top)
and CAVE WETA 20mm

These wetas have small bodies and long fragile legs. The hind legs are finely spined. The Bush Weta lives mainly in crevices in the bush and the Cave Weta is found mostly in caves. If a young weta loses a leg, the limb will regenerate, though it will always be smaller than the other legs.

Some wetas make a distinct sound by scraping the hind legs against the side of the body. The antennae and legs are easily damaged, and if lost when the wetas are young they can regrow.

Weta is a Maori name and is commonly applied to most of the weta family. Historical records show that the giant weta was called Weta Punga, and the cave wetas were not called "weta" at all. Cave wetas were greatly feared and some names associate them with the underworld: Tokoriro, Tirotiro, Po Ariki, Hine-nui-te-po, Taepo (devil), Nuku Whenua, Puruhi-o-te-whenua.

Many species of the weta are endangered and are specially protected.

CICADAS

Several species of cicadas are found in most places in New Zealand. The large wing-clapping cicadas and the green cicadas are probably the best known. They welcome in the autumn with their shrill chattering song.

After mating, the female cicada drills a series of holes with her egg-laying tube in a plant or tree in which she deposits her eggs. Females of the wing-clapping cicadas make their egg holes in a herringbone pattern, while females of other groups make them in straight lines. When the eggs hatch the nymphs fall to the ground and dig their way into the soil. They have large claw-like forelegs which they use for digging and holding on to roots of the plants and trees on which they feed.

CICADA SKIN This is the skin left by a cicada nymph, when it hatched as a mature adult, above ground.

The feeding period is at least three years. Like other insects they moult several times as they develop. When the feeding period is over the nymphs climb above the soil surface and on to the nearest available support to which they cling while the pupal skin splits and the newly-formed cicada emerges.

The male has its own "song" to attract a mate. It does not have vocal chords, but instead a special pocket or tympanum in the abdomen (under the plates near its hind legs), which is the song chamber. The song is made by pulling and releasing a ribbed membrane, sometimes at over 200 times a minute.

Among the Maori names for the cicada are Tatarakihi, Kihikihi, Kikihitara, Kikitara and Paeke depending on the area where the cicada is found. The nymph is called Ngengeti or Matua Kihikihi.

MOTTLED STAG BEETLE
Ceratognathus irroratus
This male beetle gains its name from the delightful stag-like antennae.

BEETLES

Beetles belong to the Order Coleoptera which translated means "sheath wings". There are more than 4000 different kinds in New Zealand. They are found all the year round but most often in summer. Beetle habitats include indoors and out, on the beaches, in the fields and in the mountains. For some a pond or stream is home and these are the water beetles. The largest of our beetles is the Huhu beetle which measures 40mm and one of the smallest is the tiny Rice Weevil which measures 2.5mm.

Some of them, like the wood borer, are responsible for causing much damage. Others, like the Ladybird beetle, do a lot of good. Ladybird larvae and adults feed on aphids and scale insects, or on the fungus that grows on plants.

Adult beetles lay their eggs on, or near, a source of food for the larvae. When the larvae hatch they do not look like beetles. The larvae are grubs and spend this stage of their life living inside logs or plants, while some live in the soil. Other beetle larvae have legs and are quite mobile, living on plants or trees. Some are in the larval stage for a few weeks while others are in this stage for several months.

Beetle larvae moult several times in the course of their development before pupating. When the pupal skin finally splits, the insect, in the form we recognise as a beetle, emerges.

The grub of the Huhu beetle lives in rotting wood and is delicious to eat. When at the most edible stage in its development it is known as Tataka. The grub is also used as fishing bait, and in this context is called Tui Huhu.

CHAFER BEETLES Top left

There are many species of chafer beetles. *Stethaspis longicornis* 21mm, is one of the largest and is found in scrub and bush areas. Some chafer beetles are nocturnal, and some fly in the daytime. They are all clumsy fliers, and the nocturnal species make a loud droning noise. They sometimes accidentally enter buildings attracted by the lights.

DEVILS COACHHORSE
Creophilus oculatas 20mm

A Rove Beetle, this predator feeds on other insects, and will attack flies when they are laying their eggs. Middle left

LADYBIRD BEETLES

The yellow ladybird *Illeil gabula*, it is a fungus eater. The red ladybird is the Variable ladybird *Coleophora inaequalis*, a predator. The adults and larvae feed on aphids and mites. Bottom left

LEMON TREE BORER
Oemona hirta 25mm Top right

The beetle larvae make tunnels inside citrus trees. They live and feed in these tunnels until they pupate. Eventually the trees die. The adult flies at night, and is attracted by lights. This is a Longhorn beetle.

HUHU BEETLE Bottom right
Prionoplus reticularis 40mm

The larvae of this Longhorn beetle feeds inside rotting logs. These beetles will squeak if handled.

ELEPHANT WEEVIL
Rhyncodes urses 25mm

One of the larger weevils, its larvae bore into beech trees and live in tunnels in the living tree.

WEEVILS

There are many species of weevils in New Zealand which do considerable damage. Some bore holes in the trunks and seeds of shrubs and trees. Others live in stored dry cereals and flour and spoil these products. Like beetles they belong to the Order Coleoptera, meaning "sheath wings".

Most of them are plump and their legs are not particularly long, but weevils tend to stand erect from the surface and "march". They have a long nose or "rostrum", on which are the antennae which stick out almost at right angles. If disturbed, some weevils can fold this long nose under the body, fold their antennae and fall to the ground. This posture is a perfect disguise. The weevil may lie still "playing possum" for several hours.

The Giraffe Weevil is quite distinctive. The male is long and slender and has a very long "rostrum", nearly as long as its body, with antennae at the end. The female Giraffe Weevil is short and plump, with her antennae about half way down her nose. This enables her to bore holes with her nose in which to deposit her eggs, without damaging the antennae. These weevils are found on Karaka and other native trees.

Maori names for the Giraffe Weevil are Tuwhaipapa and Tuwhaitara. The Giraffe Weevil is likened to a war canoe with its long, decorated bow wands (Ihiihi).

COCKROACHES

There are a number of species of cockroach in New Zealand. Some, like the Black Cockroach, are natives; others have been introduced accidentally. The German and American cockroaches may have arrived with European explorers as early as 1769. The Australian cockroach arrived more recently. Unlike the native cockroaches, which are happy to remain outside in their natural habitat, the introduced species are a real pest. They invade houses and industrial areas, where they may pollute and destroy food products. Cockroaches are creatures of the night and are such swift runners that they easily evade capture. They carry their eggs in an egg-case for some time until it is deposited. Only some of the species have wings.

The introduced species are: German Cockroach *Blattella germanica* 16mm; Australian Cockroach *Periplaneta australasiae* 20mm; American Cockroach *Periplaneta americana* 40mm.

Maori names for the native black stinkroach include Kereru, Kekereru and Kekerengu. Introduced cockroaches are usually called Kokoroiho after the common English name, cockroach.

BLACK COCKROACH 21mm
Platyzosteria novaeseelandiae

This is a native cockroach, nocturnal and wingless. It lives outdoors, hiding during the day under logs or rocks. If this insect is threatened it is able to produce an evil-smelling fluid, and is sometimes called Stinkroach.

BUGS

Insects are often called "Bugs", but this is a common mistake. Bugs are only those insects which suck up sap from plants or body fluid from other insects.

They are classified as Hemiptera, which can be translated as "half wing". They include aphids, cicadas, vegetable bugs, shield bugs, waterboatmen, mealy bugs and scale insects. Each has a long stylet or beak which is used to pierce food.

The two bugs that are probably seen most often are the vegetable bugs, one an introduced insect and the other a native insect. All stages of Green Vegetable Bug do considerable damage to vegetables such as tomatoes and beans. The adult females lay their eggs on these plants. These bugs do not have a larval stage or pupal stage; they have nymphs similar to adults in shape. The nymphs of both species are black with red and white speckles. The adult of the introduced vegetable bug is identifiable by the three white spots on its back.

Other true shield bugs in the garden, which are brown with light brown markings, are called Soldier Bugs. These bugs feed on other insects including monarch butterfly caterpillars.

The Passionvine Hopper is another common bug. The nymph of this insect is a little light-brown and white-speckled hopper with a plume of hairs at the end of the body.

Another frequently seen bug is the blue-green Torpedo bug. This also is a hopper, and if disturbed will jump quickly away.

There are no Maori names for this group. Insects in general are called Ngarara or Ngangara. Hirini Melbourne calls the family of insects in his songs, "Te Aitanga-a-Pepeke," the offspring of Hine Peke, who was a grandchild of Tane.

PASSIONVINE HOPPER
Scolypoppa australis 7mm

The nymphs gather on the food plant. With pale bodies and bushy tails, they are not at all like the adult, seen opposite.

GREEN VEGETABLE BUG
Nezara viridula 16mm

The nymph of the Vegetable Bug is wingless and black with pale yellow and red spots. Nymphs and adults are a pest on fruit and vegetables, sucking the sap. The resulting damage allows disease to attack the plant.
Nymph opposite, adult on right.

TORPEDO BUG
Siphanta acuta 10mm

This tiny hopper feeds on sap, and sometimes congregates in large numbers. The wings are held over the body like a tent. If disturbed it will spring away with a soft click and fly to the nearest support. This hopper is also found in blue and yellow colour forms.

PASSIONVINE HOPPER
Scolypoppa australis 10mm

The adult hopper has wings so transparent that the plant it rests on shows through the patterning. If disturbed the insect will spring into the air with a click sound then alight again.

MEALY BUGS, APHIDS and SCALES

These insects are all very small but can cause considerable damage to plants and trees. They all feed by sucking the sap from the plants. This has a bad effect on growth and may also allow disease to develop.

An example is the blight which occurs on Manuka. This is a black fungus which develops on the sugary substance exuded by the Manuka Blight Scale.

An interesting fact is that ants may "farm" mealy bugs and aphids so that they can feed on the sugary substance they produce.

Mealy bugs lay eggs. The aphid can produce eggs or live young depending on the conditions. Scales generally lay eggs.

Top: Aphids on rose bud.
Middle: Chinese wax scale *Ceroplastes sinensis* on leaf.
Bottom: Long-tailed mealy bug *Pseudococcus longispinus*.

EARWIGS

There are several species of native earwig, all wingless, and some introduced winged species. The most common earwig is the winged European Earwig.

Earwigs have claspers, or forceps, at the tail end of the body and these are used to grip prey. They feed on other insects (including earwigs), and on flowers and fruit.

The winged species have very large hind wings. These are pleated and folded up to forty times until they are small enough to be covered by the hard outer wing.

Maori names for earwigs include Mata, Hiore Kakati, Katikati and Matua Weri.

EARWIGS 20mm

Earwigs gain their name from the shape of their wings (the name earwigs is a corruption of "earwings"). Earwigs live on plants and flowers, feeding on pollen, fruit, flower buds and other insects. They sometimes live in debris on the soil and are sometimes found beneath the soil surface.

ANTS

There are 33 species of ants in New Zealand; 11 of them are natives. Depending on the species they can be red, green or black but black ants are the ones found most often in houses. Ants live together in large communities and are dependent on that community for their survival. Their nests are a series of tunnels and chambers in which they care for their eggs and feed the larvae when the eggs hatch.

There are different sorts or castes of ants; winged males and females, queens, soldiers and workers. In the colony, the workers form the largest group.

When atmospheric conditions are right, winged males and females are produced and leave the nest. They fly in vast swarms and mate in flight. During the process many ants may fall to the ground and die. The surviving females return to the ground and then each female looks for a place to start a new colony.

This new queen lays her eggs and feeds the emerging larvae herself. When these pupate and hatch they will be the nucleus of a new colony. As soon as the new workers emerge, they take over the responsibilities of feeding, building and defending the colony.

Maori names for ants vary from district to district and include Popokorua, Upokorua, Popokoriki, Rororo and Torotoro.

ANTS

Ants in the sugar bowl are a familiar sight. They are a nuisance outdoors but a real pest indoors where they invade even the most carefully-closed food containers. Yet, as a scavenger, they are most useful. Like flies they play an important part in the decomposition of waste matter.

HONEY BEE
Apis mellifera

These bees fly from flower to flower collecting honey and pollen. While collecting food for themselves they are helping nature by pollinating the flowers as they fly from one to the other.

BUMBLE BEE
Bombas terrestris

These fat furry bees are found in sunny locations, flying from flower to flower collecting nectar and pollen, pollinating the flowers as they go on their way. They have bags on their hind legs which they fill until they seem almost too heavy to fly. They are not easily angered but if persistently annoyed will sting. The sting is not barbed and is not left in the wound.

BEES

There are several species of native bees in New Zealand, but they do not build communal hives. They live singly in tunnels or burrows in sand or soil. Native bees collect pollen and nectar and store the pollen in the tunnels to feed the larvae when the eggs hatch. Maori class the native bees with flies and call them Ngaro, Rango or Rako depending on the dialect of the speaker.

The honey bees and bumble bees were introduced into New Zealand. Being larger and with longer tongues, and more furry than the native bees, they are better pollinators of introduced crops and it was for this reason that they were introduced. Maori first called the introduced bees Ngaro Miere, but now the name Pi Honi is more widely used.

The introduced bees are known as "social" insects because they live in large groups. Within the "social" community each group of insects has its own duties. The queen lays eggs, the worker bees collect food and feed the young. They also build and repair the nest and defend their domain. The male's or drone's only task is to mate with the queen bee. All summer the workers are busy collecting food both for immediate use and to store for the winter. When the winter comes the bees only leave the hive on warm days, venturing forth to find food.

The bumble bees are much larger than the honey bee, and very "furry". They build less complex nests than the honey bee. Their queen broods the first batch of eggs herself. Then, when the first brood of workers mature, they take on this care. At the onset of winter most of the bumble bees die. A few young fertile females survive however, to start new colonies in the spring. Bees are from the Order Hymenoptera; this can be translated as "membraneous wings".

ASIAN PAPER WASP
Polistes chinensis 12mm

These wasps build a mushroom shaped nest, and are fierce guardians. If approached they will immediately take a threatening stance, facing the intruder, wings at the ready to attack. If further disturbed they will attack, like miniature torpedoes, stinging then retreating to the nest ready to attack again. They are a recent immigrant to New Zealand and a nuisance in the garden.

Ancistroceros gazella 10mm
(No common name)

Easily mistaken for the Asian Paper Wasp, this wasp is a predator. The female builds a single-cell nest, carries a caterpillar to the cell as food for the emerging larva, then lays an egg in the cell and seals it.

WASPS

The best known wasp is probably the German Wasp, accidentally introduced from Europe in 1944. First found in Hamilton, it has spread through New Zealand. The German Wasp is the most savage insect we have. It will attack and kill for food a wide range of other insects, including cicadas and flies, spiders and many caterpillars. Wasps also feed on fish and meat, fruit and vegetables, juice and nectar. They will fearlessly attack humans if disturbed.

Because the sting of the wasp is smooth-sided, it can be withdrawn from its victim, enabling the wasp to sting repeatedly. This is unlike the bee, which has a barbed sting which cannot be withdrawn without mortally injuring itself. The only successful insect predator of the wasp appears to be the Praying Mantis.

Some wasps are "social" insects, and have a life style similar to that of bees. The German Wasp will search for a hole in a bank, or a crevice in a building or a tree, in which to construct a nest. Other wasps seek a protected spot and build a nest rather like an upside-down mushroom.

Mason Wasps are an example of those which build a solitary nest. They construct a series of cells. In each one they carefully place an egg and food for the larva.

Yet another group, the Ichneumonid wasps lay their eggs inside some caterpillars. When these eggs hatch the larvae feed inside and eventually kill their host. Ichneumonid wasps do not sting people unless provoked.

In Northland, a common Maori name for the German Wasp is Pi Waikato. The Red Paper Wasp is Pi Wero.

GERMAN WASP

Vespula germanica

This wasp is dangerous. It will attack if the nest is disturbed and can sting repeatedly. Unfortunately this wasp includes butterfly caterpillars in its diet.

ICHNUEMONID WASP

These wasps lay their eggs in caterpillars. When the larvae hatch they feed on the live caterpillar, eventually killing it. These wasps do not sting people.

Trissolcus bassalis

This tiny wasp lays its egg in the egg capsule of the Green Vegetable Bug. When the wasp larvae emerges from its own egg it feeds on the Vegetable Bug egg, then pupates. When the wasp is fully matured it cuts a hole in the Vegetable Bug egg case and climbs out.

MASON WASP

Pison spinolae 20mm

The Mason Wasp is often the source of the unidentified "zizzing" noise heard around buildings. The wasp builds a single-cell nest and lays her egg. Before sealing the cell she puts in an insect as food for the emerging larvae.

HOW TO STUDY INSECTS

To study an insect is both fascinating and relatively easy. First decide which insects you would like to observe. This decision will govern the choice of food, the sort of habitat you provide and size of container.

Use a clear plastic food container or make a cage from a wooden seed box. Seal the bottom of the tray to prevent the insect escaping. Place wire hoops at each end, cover with netting, as in the sketch. Put some absorbent material, such as sand or soil on the bottom, and arrange some sticks or bark to make it like the natural habitat of the insect. Remember to create a hiding place for those insects that like the dark.

If using a plastic container, cover the top with a piece of netting instead of a lid. This will allow for better air circulation and helps prevent mould growing in the container.

A good supply of fresh food and clean water is essential. The water should be in a container that allows the insect to drink but not drown. Be careful not to leave the cage in direct sunlight. It is also important to keep the cage free from old and mouldy food. Insects kept in dirty containers can become sick and die.

Once you have created the right conditions, the insect will then do the work and you will be able to watch it grow and moult, pupate and emerge as an adult. Insects generally do not feed while they are moulting. This can be a sign for you to watch carefully so as not to miss seeing the insect emerge. When insects moult, their bodies are very soft and at first very pale in colour. It may be several hours before their wings and bodies harden and their colours form. When an insect emerges do not attempt to handle it, for at this stage it is easily damaged.

Do not keep an insect too long once it reaches the adult stage. Take it into the garden, or preferably the habitat where it was found, and release it. Then the insect can find a mate and the reproductive process will begin again.

The container breeding method is suitable for large insects that cannot escape through the netting. Finer mesh materials, like nylon stocking, may help hold tiny insects. Extreme care must be taken with those insects that bite or sting.

Macromastix viridis 20mm

This large, green crane fly is usually found near the bush.

FLIES

Few people will list flies among their favourite insects, yet seen close up they are very colourful with their huge compound eyes and shimmering wings. Flies belong to the Order Diptera which means "one pair of wings". People associate flies with dirt and disease, but they also play an important role in the breaking down of waste matter. The most common flies are the house and blow flies. Other flies include crane flies, sandflies, fruitflies and mosquitoes. All have mouthparts equipped either for biting or sucking.

Flies all have the same three stages in their development. The larva is a legless "maggot" which feeds on living tissue or decaying matter. Larvae moult several times in the course of their development, then pupate. The pupal case of flies like house and blow flies is barrel-shaped, and quite hard, and is called a puparium.

Maori call the stout-bodied house flies, and blow flies Ngaro, Rango or Rako, according to the district. The long-legged crane flies and mosquitoes are called Waeroa in the north. Maggots are called Iroiro or Kutukutu. In historical records, the native "blue-bottle" blow fly is called Rango Pango, and the golden-haired blow fly from Australia is called Rango Tuamaro.

BLUE BLOW FLY
Calliphora quadrimaculata
13mm

The Blue Blow Fly will enter buildings and spoil unprotected food. It has huge compound eyes which give the fly all-round vision, helping it to evade the intrepid fly swatter.

GREEN BOTTLE FLY
Lucillia sericata 13mm

These flies rarely come inside. They are sometimes copper coloured, but always have a characteristic metallic sheen. They cause "fly-strike" on sheep.

CRANE FLIES

There are around 600 species of these flies in New Zealand. The adults are usually found outside at night, attracted to lights. They are very fragile and easily lose their long legs if handled. The larvae are called Leatherjackets. These leathery, shiny grubs are found in the soil. The Daddy-long-legs insect is a crane fly.

DRONE FLY *Eristalis tenax*

This is a fly that looks like a bee (note it has only one pair of wings). This fly feeds on pollen and, because the pollen sticks to its "furry" body, it is a pollinator and a beneficial insect. The larvae live in wet areas and are called Rat-tailed Maggots.

Index of Common Names

Antlion	22
Ants	23
Aphids	22
Auckland Weta	12
Bees	3, 25
Beetles	15, 16, 17
Black Cockroach	19
Black Field Cricket	11
Blue Blow Fly	2, 31
Bumble Bees	25
Bush Weta	13
Cave Weta	12, 13
Chafer Beetles	16
Cicadas	14
Cockroaches	19
Common Mantis	6, 7
Crane Flies	30, 31
Crickets	10, 11
Daddy-Long-Legs	31
Devil's Coachhorse	16
Earwigs	23
Elephant Weevil	18
Flies	5, 30, 31
Fruitflies	30
German Wasps	26, 27
Giraffe Weevil	18
Grasshoppers	10, 11
Green Bottle Fly	31
Green Vegetable Bug	20, 21
Honey Bee	24, 25
House Fly	5
Huhu Beetle	15, 16
Ichneumonid Wasps	27
Katydids	10, 11
Ladybirds	3, 15, 16
Lemon Tree Borer	17
Longhorn Beetle	2
Mantids	Covers, 2, 6, 7
Mason Wasp	26, 27
Mealy Bugs	22
Mosquitoes	30
Musca domestica	5
Passionvine Hopper	20, 21
Praying Mantis	Covers, 2, 6, 7
Scale Insects	22
Soldier Bugs	20
South African Mantis	6, 7
Stick Insects	8, 9
Torpedo Bug	21
Trissolcus bassalis	21
Wasps	26, 27
Weevils	18
Wetas	12, 13

Further Reading

D. Miller, *Common Insects in New Zealand*, revised A. Walker, A.H. & A.W. Reed, 1984.

R. Sharell, *New Zealand Insects and their story*. Collins, 1971.

R.R. Forster & L.M. Forster, *Small Land Animals of New Zealand*, reprinted John McIndoe, 1980.

Standard Names for Common Insects in New Zealand, Bulletin 4, Entomological Society of New Zealand, 1977.

Acknowledgments

I would like to thank all the friends who have assisted me in identifying the insects used in the photographs, Dr Peter Maddison for his encouragement to write this book, and Wendy Pond for providing the Maori names. These names were provided with linguistic macrons to indicate pronunciation but are published here in their popular form because of the general nature of this book.